To:

From:

I hope that this book helps you to plan and achieve your goals

M Boxall

First published by Mark Boxall in 2021

Copyright © Mark Boxall, 2021

All rights reserved. No part of this publication may be reproduced, stored, or transmitted in any form or by any means, electronic, mechanical, photocopying, recording, scanning, or otherwise without written permission from the publisher. It is illegal to copy this book, post it to a website, or distribute it by any other means without permission.

But how will I do that???

What is a Scientist?

A Scientist is someone who asks questions ...

And enjoys solving puzzles.

What does a Scientist do?

Scientists do research, find answers to difficult questions and solve problems.

They do this by Observing, Monitoring and Communicating.

Observing

Put simply, this means looking at and examining things.

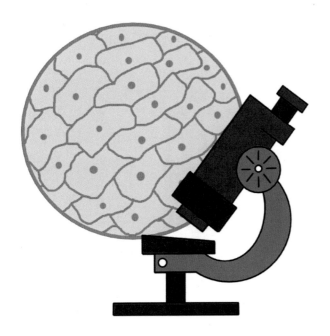

Measuring

You must record your observations as accurately as you can.

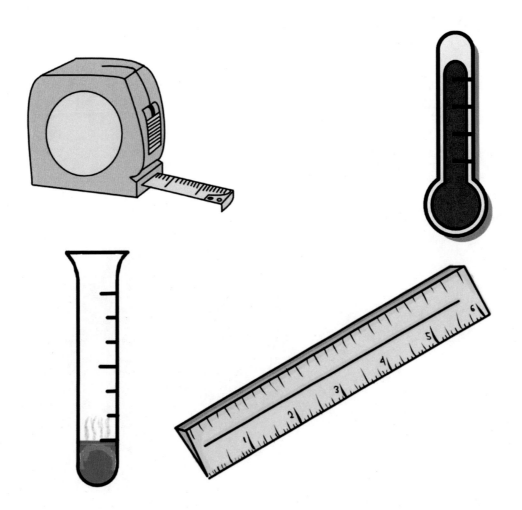

Communicating

The next step is to tell other people what you have observed and recorded and talk about your results.

There are three basic questions that scientists ask:

Why – Why does this happen?

How – How does this work?

What – What are the rules that apply?

Then you ask more questions.

How does this affect other things?

For example, what happens if I mix yellow paint with blue paint?

Keep asking questions and communicating until you have answers that everyone agrees on, and experiments that other people can do to get the same results.

This part is a bit like cooking – someone experiments with different ingredients and cooking methods, then writes a recipe so that other people can make the same dish.

Scientists work in teams and collaborate with each other so you will have to learn how to work with other people.

TOP TIP

Taking part in social activities will help you learn how to get along with other people and work as part of a team.

This can be anything from joining a chess club to playing team sports.

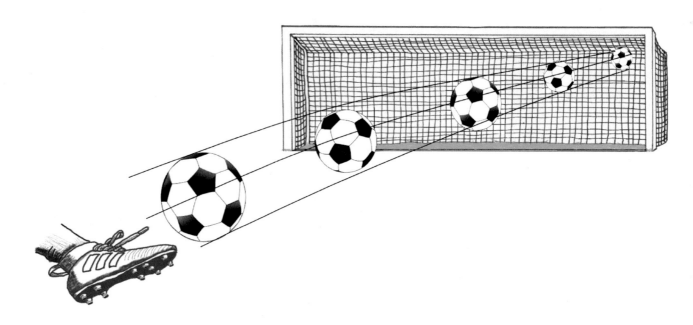

How many social activities can you think of?

Exercise is not just a good way to keep fit and healthy.

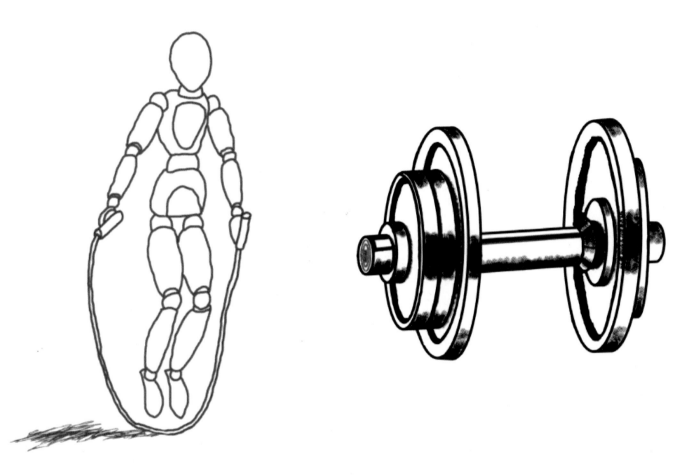

It can be an enjoyable way to meet and interact with other people.

How many different types of exercise can you think of that you can do in a group?

So now we know what a Scientist is, but what is Science?

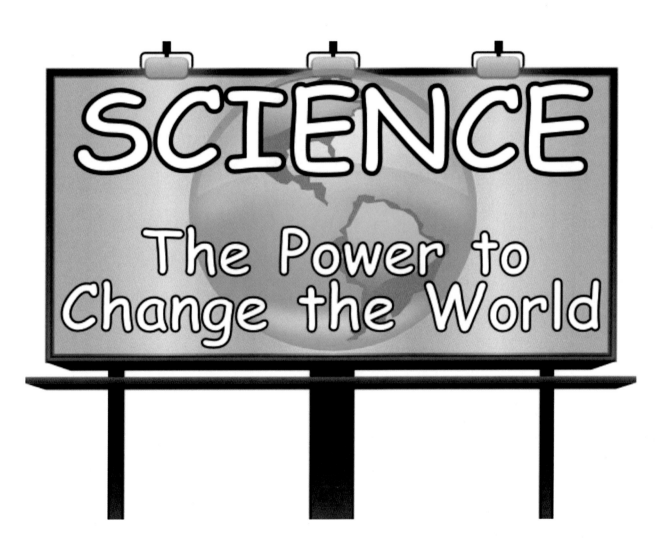

To put it simply, Science is the study of everything.

From how The Universe started,

to how people can live on Mars.

From why all the dinosaurs died,

to cleaning up the environment.

Want to know more about that interesting stone you found?

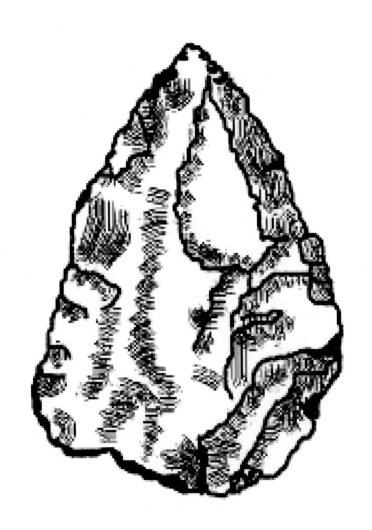

Ask a **Geologist**.

How about the pretty shell you found on the beach?

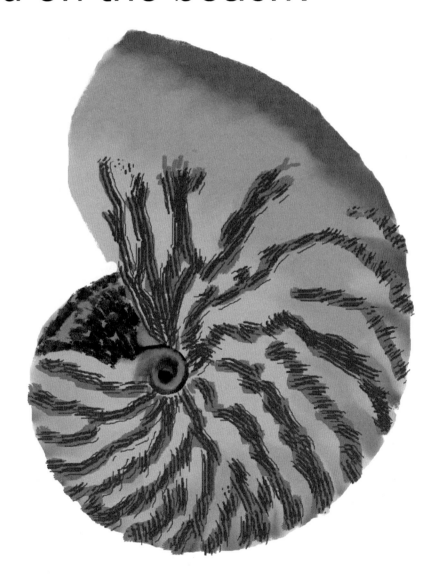

Ask a **Marine Biologist.**

Which Stars are actually Planets?

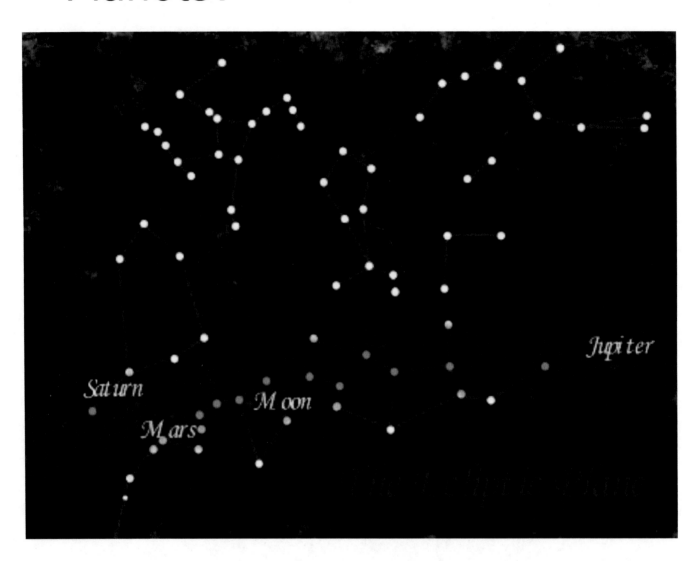

Ask an **Astronomer.**

How does my stomach work?

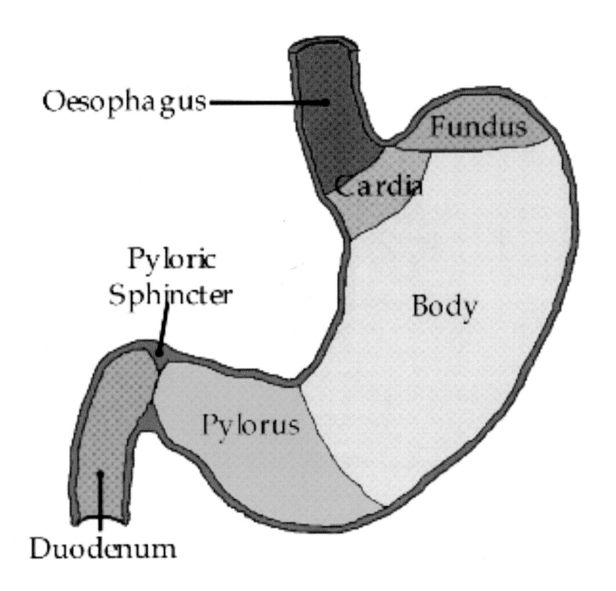

Ask a **Gastroenterologist**.

A healthy diet is very important for everyone.

A **Dietitian** knows what foods are healthy and which ones have vitamins and minerals that your body needs.

They also know about diseases and illnesses and how proper nutrition can help treat or prevent them.

Remember – to be the best you have to work the hardest

Diet

Study

Hard

Work

Exercise

Practice

Final Tip

To be a Scientist you will have to study and understand a lot of things. It is recommended that you base your studies around **STEM.**

STEM stands for:

Science

Technology

Engineering

Mathematics

Here are five interesting websites that will help you learn more about science.

www.nasa.gov/kidsclub

For everything to do with space.

www.astrostem.org

Based around STEM topics.

www.sciencekids.co.nz

The online home of science and technology for children.

www.bbc.co.uk/teach/terrific-scientific

The science page of the BBC Teach website

www.thekidshouldseethis.com

Ideal for curious kids like you.

Anyone from any background can be a Scientist.

There is a lot of effort being made to ensure that there is no bias and equal opportunities for everyone who wants to become a Scientist, so If you do all these things and keep working hard then when you grow up you will be a

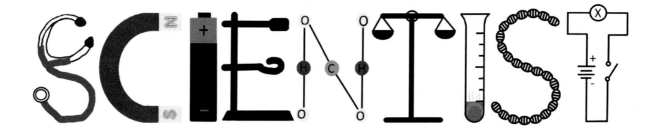

And finally...

When I grow up, I want my family to be proud of me.

This has been a Sparky B Production.

sparkybproductions.mysites.io/home

Made in the USA
Monee, IL
09 December 2022

20463095R00024